DATE DUE

Material World

SEPARATING MATERIALS

Robert Snedden

Heinemann Library
Chicago, Illinois

Designed by Celia Floyd
Originated by Dot Gradations
Printed by Wing King Tong, Hong Kong

05 04 03 02 01
10 9 8 7 6 5 4 3 2 1

Library of Congress Cataloging-in-Publication Data
Snedden, Robert.
  Separating materials / Robert Snedden.
    p. cm. -- (Material world)
  Includes bibliographical references and index.
  ISBN 1-58810-070-7 (library binding)
  1. Separation (Technology)--Juvenile literature. I. Title.

TP156.S45 S63 2001
660'.2842--dc21

                                          00-061344

Acknowledgments
The author and publishers are grateful to the following for permission to
reproduce copyright material: Andrew Lambert, pp. 12, 13; Bruce Coleman
Collection/Erik Bjurstrom, p. 8; FLPA/M. Rose, p. 9; FLPA/M. J. Thomas, p. 28;
Heinemann/Trevor Clifford, p. 15; Hulton Getty, p. 19; Robert Harding Picture
Library, p. 22; Science Photo Library/David Parker, p. 5; Science Photo
Library/Charles D. Winters, p. 7; Rosenfeld Images Ltd, p. 14; Rosenfeld Images
Ltd/Jerry Mason, p. 26; Rosenfeld Images Ltd/Alex Bartel, p. 29; Tony Stone
Images/Phil Degginger, p. 4; Tony Stone Images/Rich Iwasaki, p. 11; Tony Stone
Images/David Woodfall, p. 18; Tony Stone Images/Keith Wood, p. 20.

Cover photograph: Science Photo Library:/Alex Bartel.

Every effort has been made to contact copyright holders of any material
reproduced in this book. Any omissions will be rectified in subsequent printings if
notice is given to the publisher.

Some words are shown in bold, **like this**. You can find out what
they mean by looking in the glossary.

Contents

Mixed Materials

We use a number of different materials in an endless variety of ways. Most materials have to be changed in some way before they can be used. Some materials are natural, such as stone and wood. These can be used almost as they are and may only need to be shaped. However, most useful materials are usually found with other materials that are not so useful. Others, such as plastics, have to be manufactured from **raw materials.** Here are a few examples of useful materials that have to be separated from other materials before they can be used.

Metals

Few metals are found in a pure form. They react with other **elements** to form **compounds** that are not as useful to us. Before we can use the metals we have to find ways to separate them from the rocks, called **ores,** in which they are found.

In a foundry, molten metal is separated from some of the impurities found with it.

Petroleum

Petroleum, or crude oil, is a complex **mixture** of many different oils and other compounds. Each of the oils has its own characteristics and its own uses, but before they can be put to use they have to be separated. This is the job of oil **refineries.**

Water

The water that comes from your tap is a mixture, having many different chemicals **dissolved** in it. Of course, the water companies have to ensure that any potentially harmful substances have been removed from it before it reaches you.

This book is about the different ways materials can be mixed together—and some of the methods that have been developed to "unmix" them. All of these techniques depend on us having an understanding of the properties of the materials we are trying to separate. Knowing how substances behave together helps us to determine ways of obtaining the materials we want. For instance, to separate the materials we need from the ones we do not want, we have to know how it is that they have come together. This knowledge helps scientists to develop efficient ways of separating materials.

Desalination plants, like this one, turn salt water into fresh water for drinking and irrigation.

Compounds and Mixtures

All matter is made up of tiny **particles** called **atoms**. An **element** is a substance that is made up of just one kind of atom. An atom of one element is different from those of any other element. Examples of elements include: oxygen and nitrogen, which are gases in the air we breathe; copper, a metal used for conducting electricity; and carbon, the **graphite** used in pencils.

Atoms of different elements can combine in a variety of ways to form millions of **compounds.** A compound is a substance that contains more than one kind of atom. Examples of compounds include: common salt, which contains atoms of sodium and chlorine; and carbon dioxide, which has atoms of carbon and oxygen. The atoms are joined together by chemical **bonds** and cannot be separated again by physical means, such as **distillation,** or mechanical means, such as **filtering.**

Atoms of different elements can join together to form compounds, like when carbon combines with oxygen to form carbon dioxide.

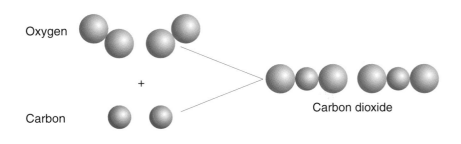

Oxygen

+

Carbon

Carbon dioxide

In proportion

In a compound, the atoms always combine in the same proportions. For example, water always has two atoms of hydrogen to every one atom of oxygen. Sometimes atoms of the same elements combine in different proportions to produce different compounds. Atoms of carbon and hydrogen can combine to form methane, the main component in natural gas. But there are also thousands of other compounds that contain only carbon and hydrogen, such as gasoline and kerosene.

Compounds may be solids, liquids, or gases. The properties of a compound have little in common with the properties of the elements that make it up. Here are some examples:

- *hydrogen and oxygen are both gases under normal conditions, but when combined they form water, which is a liquid at normal temperatures*

- *sodium, a soft metal, combines with chlorine, a yellowish-green poisonous gas, to form sodium chloride which is common salt, a hard, white, crystalline solid*

- *carbon, a solid, and oxygen, a gas, combine to form carbon dioxide, a gas.*

A carbonated drink is a mixture of water, sugar, flavorings, and carbon dioxide.

Mixtures

Many substances that contain atoms from more than one element are **mixtures,** not compounds. The different substances in a mixture do not react together to form new substances as they do in a compound. A mixture can be made up of different elements, such as iron filings mixed with carbon soot, or of different compounds, such as the water, sugar, and carbon dioxide gas in a carbonated drink. Usually we can find a physical way, like filtering or distilling, to separate the different parts of a mixture. Compounds have to be separated chemically.

Solutions

A **solution** is a **mixture** of two or more different substances that cannot be separated by a mechanical method, such as **filtering.** Liquid solutions are the most common. These result when one or more substances—liquid, solid, or gas—are **dissolved** in a liquid. The dissolved substances are called **solutes** and the liquid in which they are dissolved is a **solvent.** An example of a liquid solution is sugar (the solute) dissolved in water (the solvent). Gases and solids that dissolve in a liquid are said to be soluble.

The ability of a substance to dissolve in another substance is called its solubility. The solubility of most solids depends on the chemical properties of the solute and the solvent and on the temperature of the liquid solution. A given **volume** of a solvent at a particular temperature can dissolve only a certain amount of solute. You can see this if you go on adding sugar to coffee—eventually there will be undissolved sugar at the bottom of the cup. For solutions involving gases, solubility also depends on **pressure.**

Fish use their gills to breathe oxygen that is dissolved in the water.

8

Miscible liquids

Two liquids that have the ability to form a solution are said to be miscible. Water and alcohol, for example, are completely miscible and will readily form a solution. Water and oil, on the other hand, do not mix well at all. Whether or not two liquids can form a solution together depends on the chemical properties of each liquid. Physical conditions such as temperature and atmospheric pressure can also play a part.

Solid solutions

When liquid solutions freeze, the components do not separate so the result is a "solid solution." A metal **alloy** is an example of a solid solution. It is a combination of two or more metals, such as the mixture of melted copper and zinc that forms brass.

Gas solutions

The air that we breathe can be thought of as a gaseous solution, formed from a mixture of nitrogen and oxygen, plus smaller amounts of other gases such as argon and carbon dioxide. Physical conditions, like temperature, do not affect the ability of gases to form solutions.

Solution caves are formed when surface water trickles down through tiny cracks in limestone or similar rocks and slowly dissolves the rock over thousands of years.

Suspensions and Emulsions

A **suspension** is a liquid or gas that contains small **particles** of another material. There are several types of suspensions. These include:

- a solid in a gas, such as particles of smoke drifting in the air;
- a liquid in a gas, such as fog, which is water **vapor** in the air;
- a solid in a liquid, such as the cloudiness seen in a muddy pond;
- a gas in a liquid, such as the foam used in fire extinguishers;
- a liquid in a liquid, such as water-based paints.

A suspension that contains extremely small particles is called a **colloid.** The particles in many colloids can only be seen with the aid of a microscope. The tiny globules of suspended fat in milk are an example of a common colloid.

The foam used in fire extinguishers is a suspension of a gas, carbon dioxide, in a liquid, water.

Brownian motion

The particles in a suspension are constantly moving. This is because they are always being buffeted by the fast-moving particles of the liquid or gas they are suspended in. The rapid movement of the particles that results from these collisions is called Brownian motion, after the Scottish scientist Robert Brown who first described it.

Suspensions and solutions

You may have noticed a beam of sunlight shining into a room through a gap in the curtains. You can see the beam because tiny suspended particles of dust in the air reflect and scatter the light. When a beam of light is shone through a colloid, like the fine dust in the air, the path of the beam becomes clearly visible. This does not happen if light is shone through a **solution** because the particles are too small to scatter light.

Using an appropriately sized **filter,** it is possible to separate out the particles in a suspension. However, no filter is fine enough to separate out the **molecules** in a solution.

Emulsions

An emulsion is a type of colloid that is produced when one liquid is evenly dispersed in another. It is not a solution—the two liquids do not dissolve in each other—but tiny drops of one liquid are suspended in the other liquid. Some common substances, such as lotions and paints, are emulsions. Milk is an emulsion of butterfat in water.

Smog, which can be found in many large cities, is a suspension of pollutants in the water vapor in the air.

Sieves and Filters

A **filter** is a device for separating unwanted substances from liquids or gases in a **mixture**. Filters that remove solid **particles** or other impurities from liquids or gases can be made from **porous** materials, such as paper, cloth, charcoal, and porcelain. The filter has small holes that are only large enough to let the liquid **molecules** pass through. The solid particles are held back. The liquid that passes through the filter is called the **filtrate**. The solid that is left behind is called the **residue**.

Sieving is a way of separating larger particles from smaller ones, or from a liquid.

A sieve is a type of filter that is used to separate out larger particles. When you drain cooked rice through a sieve you are filtering out the rice from the water. In this case you eat the residue. A tea strainer, used to trap tea leaves, is another example of filtration—but in this instance, it is the filtrate that we are interested in.

Filters at work

Internal combustion engines use various types of filters to remove impurities from air, lubricating oils, and the fuel they use. It is important that this is done so that the engine will run smoothly. Dry-paper filters on **carburetors** remove impurities from the air before they enter the engine. Most oil filters are also made of paper. Most water-purifying filters use either **ceramic** or glass fibers to trap unwanted dirt and bugs. Ceramic is made up of millions of honeycombed pores large enough to let water pass through, but small enough to trap bacteria. Fiberglass filters have the advantage of being stronger than ceramics.

Filters are used in car engines to trap impurities that might prevent the engine from running smoothly.

Try it!

Can filters separate all residue from a mixture?

You will need
water
soil
two jars
a coffee filter

1. Pour water into one of the jars and then stir a little soil into it to make a cloudy **suspension.**
2. Put the coffee filter into the top of the clean jar. Slowly and carefully pour the muddy water from the first jar through the filter paper, into the second jar.

The coffee filter will trap a lot of the soil (the residue). The water in the second jar (the filtrate) should look cleaner. However, it is not clean enough to drink. The water will probably still look fairly cloudy as you will not have been able to remove the finest soil particles—or the multitudes of bacteria from the soil.

Clean Water

How do we make sure that the water that comes from the tap is clean and safe?

The separating processes used in water treatment plants are similar in many ways to natural processes. Soil is a natural **filter** and materials can be removed from water as it moves through soil. Materials often settle out from water in a lake, drifting down to the bottom. This is called settling. When water **evaporates,** it is purified as the individual water **molecules** become **vapor** and drift into the atmosphere, leaving other materials behind. This process is called **distillation.**

The simplest way of cleaning up our water is by filtration and settling. At the same time, waste materials that float on top of the water can be skimmed off. Half of the pollutants in the water can be removed by these simple methods.

Waste eaters

The next stage is a biological process. Useful **microorganisms** consume most of the waste material in special tanks. Solids and microorganisms are separated from the waste water in settling tanks. Disinfectants, such as chlorine, may be added to the water to kill any remaining disease-causing organisms.

A water treatment plant is where harmful impurities are removed and water is made safe to drink.

Filtering and flocculating

There may still be small amounts of undesirable materials in the water, such as chemicals from pesticides and cleaning materials. The next stages of treatment help remove most of these materials. Filtration through **activated carbon** removes organic materials and distillation removes salts. Flocculation is a process that involves adding a chemical to water that causes suspended **particles** to clump together and settle out.

Some pollutants are extremely difficult to remove from water. If you pour some household cleaners or garden pesticides down the drain you are adding chemicals that might not be removed during treatment.

Light cleaning

In 1996, Ashok Gadgil of the Lawrence Berkeley National Laboratory in California invented a water purifier that uses ultraviolet light to remove bacteria. The ultraviolet light damages the bacteria's DNA, making them harmless. Water flows down through pipes into a tray where it is exposed to twelve seconds of ultraviolet light before it flows out again. Gadgil tested the purifier on water contaminated with E. coli bacteria and found that fewer than 1 in 100,000 of them survived. The purifier did just as well against several other germs, including typhoid and cholera. A company in Bombay is making a solar-powered version that can be used in villages in India.

Evaporation and Distillation

The **molecules** that make up all materials are always moving. They always have a certain amount of **kinetic energy.** The molecules in a liquid have more energy than the molecules in a solid. The more energy the molecules have, the faster they move. **Evaporation** occurs when the molecules of a substance have enough kinetic energy to escape from the substance's surface as **vapor.**

Energy in the form of heat can speed up evaporation. For example, puddles of rainwater dry up and disappear more rapidly if the Sun comes out after a shower. The increase in temperature increases the energy of the water molecules and they escape at a faster rate.

The **particles** in a liquid are constantly moving. Evaporation occurs when a particle has enough energy to escape from the surface of the liquid into the surrounding air.

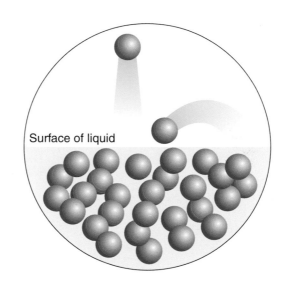

Surface of liquid

Volatility

Different substances evaporate at different rates. For example, alcohol will evaporate much faster than water at the same temperature. Substances that have fast rates of evaporation are said to be **volatile.** Solids actually evaporate too, but so slowly that the process never becomes noticeable.

Evaporation at work

Because evaporating molecules remove heat from the substance they escape from, evaporation can be used for cooling. As perspiration evaporates from your skin, it cools you down. This is why you sweat on a hot day or after exercise. Air-conditioning and refrigeration systems use evaporators to remove heat and moisture from the air and so reduce the temperature.

Distillation

Distillation is a process that separates a substance or a **mixture** of substances from a **solution** through vaporization. It can be used to separate mixtures of liquids that have different volatilities and usually involves heating the mixture and **condensing** the vapor that forms. The liquid collected in this way will be largely made up of the more volatile of the substances. Distillation is used in oil **refining** to separate out the different parts of crude oil, or petroleum. We will be looking more closely at this process on pages 20–21.

Distillation is carried out in a still. A still has three main parts: a boiler, a condenser, and a receiver. The mixture to be **vaporized** is heated in the boiler. Whichever substance in the mixture boils at the lowest temperature will be the first to turn into vapor. The vapor enters the condenser, where it cools and becomes a liquid again. The distilled liquid, called the distillate, is then collected in the receiver.

Distillation can be used to separate salt and other impurities from water.

Density and Decanting

Materials with different **densities** will separate, as you can see when oil floats on water. The materials can then be separated by decanting—pouring off the less dense material on top.

The density of a substance is found by dividing its **mass** by its **volume**. Measuring the density of a substance is one method that scientists use to identify minerals and other solids. Chemists can measure the density of a **solution** to work out the concentration of a substance in that solution.

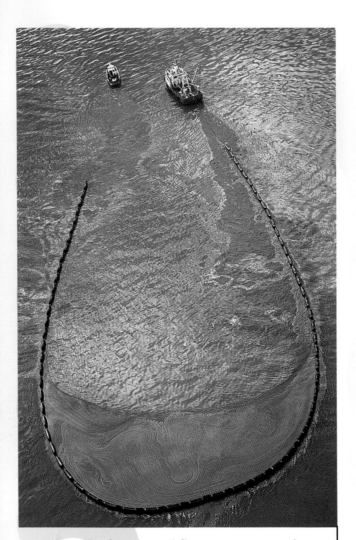

The fact that oil floats on water makes a spill slightly easier to deal with as some of the oil can be skimmed off the surface.

Try it!

How can you measure density at home?

You will need
a plastic pitcher
a kitchen scale
a variety of small objects

1. Pour enough water into the pitcher so that it would submerge any of the objects, but do not fill the pitcher to the top.
2. Note how much water is in the pitcher by reading it off the scale.
3. Place one object on the kitchen scale and note the weight.
4. Submerge the object in the water. Look to see where the water level is now. The difference equals the volume of the object. Note this.

The density of the object can be worked out by dividing the weight by the volume. Try this using different objects.

Decanting

Decanting can be used to separate out substances of different densities from a **mixture.** For example, if the **particles** in a liquid were dense enough, they could be left to sink to the bottom of a container and the liquid could be poured off leaving them behind. Milk sometimes has some of the cream it contains removed to reduce the milk's fat content. This used to be done by allowing the less dense cream to rise to the top of the milk and then skimming it off. Now, it is done using a centrifuge, a device that whirls the milk around very rapidly. The less dense cream gathers in the center where it can be removed.

Looking for gold

Gold prospectors used decanting methods to search for gold nuggets. The prospectors would use large, shallow pans to scoop up water, sand, and gravel from a river where they thought there might be gold. Then they would swirl the mixture around in the pan. Gold is a particularly dense substance and would sink to the bottom of the pan, while the muddy water could be poured out (decanted). This method is called panning.

Washing and decanting to extract gold from rocks was hard work, but sometimes the rewards made it all worthwhile.

Oil Refining

An oil refinery spreads over a large area, processing and refining hundreds of thousands of barrels of crude oil every day.

Crude oil, or petroleum, is mainly made up of a **mixture** of different **compounds** of hydrogen and carbon (hydrocarbons). The job of an oil refinery is to take this mixture and separate it out into useful products. A large refinery can process more than 200,000 barrels of oil a day.

Fractional distillation

The first stage in oil **refining** is **fractional distillation**. This separates crude oil into different parts, called fractions.

Crude oil is pumped through pipes inside a **furnace** and heated to temperatures of around 752°F (400°C). The mixture of hot gases rises through a vertical steel cylinder called a fractional distillation column. As the gases rise in the tower, they cool and **condense** at different levels.

Fractional distillation takes advantage of the fact that different compounds in the oil **vaporize** at different temperatures. For example, kerosene vaporizes between 302 and 527°F (150 and

275°C) whereas gas oil, used to make diesel fuels, vaporizes between 428 to 662°F (220 to 350°C). Residues of hydrocarbons that do not vaporize are recovered from the bottom of the tower and used for products such as asphalt.

Heavy fuel oils condense first and kerosene and gasoline condense in the middle and upper sections. At the top of the tower, gases such as methane and butane are collected.

Cracking

Scientists have developed ways of converting less useful oil fractions into more valuable ones. Cracking converts larger hydrocarbon **molecules** into smaller ones, such as gasoline. Gasoline produced by cracking is of a better quality than that produced by fractional distillation.

The large molecules can be broken down by applying great heat and **pressure** or by the use of a catalyst to speed up the cracking process. A catalyst is a substance that changes the speed of a **chemical reaction** without being changed itself. In catalytic cracking, the fractions are heated and then passed over **minerals** called zeolites. The combination of the heat and the catalysts cracks the heavy molecules into lighter ones.

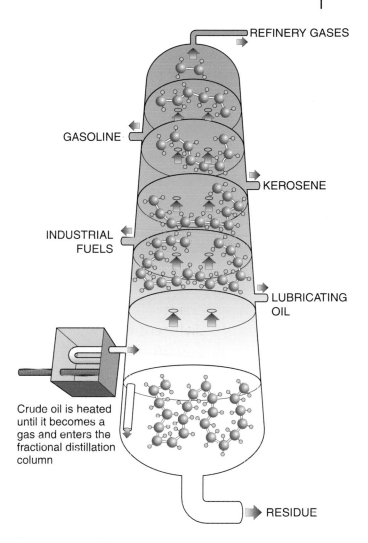

REFINERY GASES

GASOLINE

KEROSENE

INDUSTRIAL FUELS

LUBRICATING OIL

Crude oil is heated until it becomes a gas and enters the fractional distillation column

RESIDUE

Removing impurities

The most common impurities in oil fractions are sulfur compounds. These can damage machinery and are also a source of pollution when the fuels are burned. The fractions are mixed with hydrogen gas and then heated and exposed to a catalyst. The sulfur in the fractions combines with the hydrogen, forming hydrogen sulfide which is later removed by **dissolving** it in a suitable **solvent.**

Ores and Extraction

Most of the metals we use are found in the form of **ores**—the rocky material from which the metal has to be extracted. The science of separating metals from their ores and making them ready for use is called metallurgy. There are a wide variety of processes for getting metals from their ores.

Mineral dressing

One of the first stages is **mineral** dressing. This is done after the ore has been mined. The aim of the process is to remove as much of the waste material as possible from the ore. The ore is crushed and then set in motion in water with bubbles of gas. Chemicals or oils are added to the water that cause the mineral **particles** to stick to the bubbles. The minerals can then be removed in a froth. The waste materials are left behind and can be washed away. The removal of this waste reduces the amount of ore that has to be processed to extract the metal.

Roasting and smelting

Most metal ores are complex **compounds** consisting of metal **atoms** combined with oxygen, silicon, or sulfur. Unwanted materials have to be removed from the compound, leaving the metal behind. Often this is done at a high temperature. When sulfur-containing ores are heated in air, the sulfur combines with oxygen in the air forming gases that can be removed. This is called roasting.

In producing steel, iron ore, which contains iron oxide, is placed in a blast **furnace** along with limestone and coke (residue of coal after **distillation**). This is called smelting. Oxygen is blown into the blast furnace where it combines with carbon in the coke to produce carbon monoxide. The carbon monoxide then removes the oxygen from the iron oxide, combining with it to form carbon dioxide. The liquid iron left behind is called pig iron. The limestone melts into a slag that combines with phosphorus and sulfur impurities from the liquid iron. This is lighter than the iron and rises to the top of the molten metal where it can be drawn off through holes in the side of the furnace.

Leaching

Some metals can be separated from their ores by **leaching**. This involves **dissolving** the metal out of the ore using a chemical **solvent**. The metal is then taken out of the **solution** by precipitation. Precipitation occurs when two compounds react together in a solvent to produce an **insoluble** compound that comes out of solution. Gold is often separated from its ore by dissolving the gold in the ore with a solution of sodium cyanide. After the gold has dissolved it is placed in contact with metallic zinc. This causes all the gold to separate out from the solution and gather on the zinc.

Some metals can be separated from their ore by a process called leaching.

Electrolysis

Some materials are so tightly bound together that they can only be separated using electricity. Electrolysis is a process in which an **electric current** is passed through a **solution,** causing a **chemical reaction** to take place. If the solution contains a metal, electrolysis breaks up the solution so that the metal can be removed. Electrolysis will only work in **refining** substances that **conduct** electricity, such as copper and aluminum.

Electrolysis at work

Two solid electrical conductors, called **electrodes,** are placed into a liquid. These rods can be metal or **graphite.** The electrodes are connected by wires to battery terminals or to a direct current generator. The liquid must contain a substance that can carry the current and complete the electrical circuit. This is called an electrolyte. The electrodes, liquid, and the container that holds them make up an electrolytic cell.

As the current flows through the electrolytic cell, chemical changes take place at the electrodes. On the surface of the **cathode,** which is connected to the battery's negative terminal, the electrolyte combines with **electrons** supplied by the battery. This process is called reduction. At the surface of the **anode,** connected to the battery's positive terminal, the electrolyte gives up electrons. This process is called oxidation. When solutions containing **ions** of metals, such as copper and silver, are electrolyzed the metal is deposited on the cathode.

Copper is **leached** from some **ores** using sulfuric acid to produce a copper sulfate solution. The solution is placed in an electrolytic cell and an electric current is passed from a lead anode, through the solution and to a copper cathode. The copper **particles** in the solution are deposited on the copper cathode.

Electrolysis is also used to produce sodium metal by the electrolysis of molten sodium chloride (common salt). In this process, chlorine gas is produced at the anode. Both sodium metal and chlorine gas have many important industrial and chemical uses. Electrolysis is also used to produce magnesium and aluminum for industry.

Try it!

How will electrolysis effect salt **dissolved** in water?

You will need
salt
a glass of water
a 9-volt battery
two lengths of copper wire
aluminum foil
tape

1. Cut out two small squares of foil and tape them to one end of each wire.
2. Fix the other ends of the wires to the battery terminals.
3. Dissolve a spoonful of salt in the glass of water and dip the foil squares into the glass.

Bubbles of gas will start to form on the foil squares as the electric current splits the salt and water molecules.

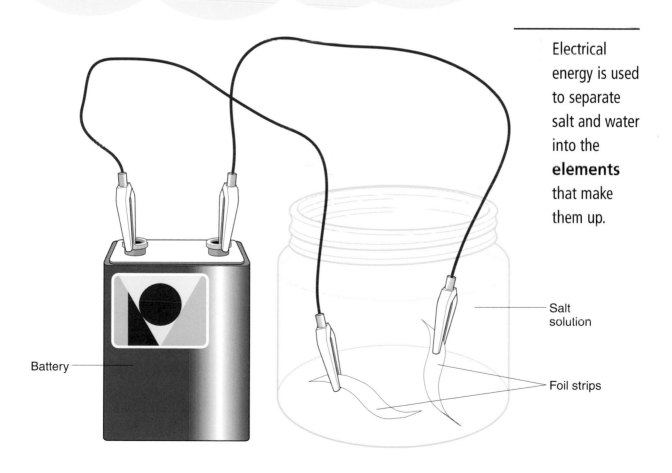

Electrical energy is used to separate salt and water into the **elements** that make them up.

Battery

Salt solution

Foil strips

Chromatography

Chromatography is a way of separating a **mixture** of substances. It is used by biologists and chemists to analyze different materials and discover what they contain. It works by **dissolving** the mixture and then drawing it through a medium, such as **filter** paper or oil, with which it will not mix. Different substances in the mixture will travel through the medium at different rates and will be separated.

Adsorption

Chromatography works mainly because of a process called adsorption. Adsorption is what happens when a substance is attracted to a solid or liquid material and forms a thin film on its surface. It is different from absorption, which involves a liquid being soaked up by something. When a mixture is to be analyzed by chromatography, it is dissolved in a liquid or a gas to allow it to move more easily across an adsorbent material, such as filter paper. As the mixture moves across the adsorbent material, the components separate out because some are more strongly adsorbed and so move slowly, while others are weakly adsorbed and so move more quickly.

Chromatography can be used to identify tiny parts of a mixture. Computers can then help identify an unknown mixture by looking for various properties of the possible ingredients.

Types of chromatography

The following chromatographic methods are commonly used in the laboratory.

Liquid column chromatography
A glass tube is filled with an adsorbent material. The mixture to be separated is dissolved in a suitable liquid and is added at the top of the column. The substances move down through the column at different speeds, separating out as they do so.

Thin layer chromatography
This process uses a thin, flat sheet of glass or other material coated with an adsorbent **film**. A drop of the mixture is placed on one end of the sheet, which is then stood on end in a shallow pool of liquid. The liquid travels up the film, carrying the mixture along with it. The substances separate from one another as they are adsorbed by the adsorbent film.

Gas chromatography
In this method, a column of adsorbent material is used to separate gases and substances that are easily converted into gases by heating. A carrier gas flows through the tube to keep things moving. The gas most often used for this is argon, which is unreactive.

Try it!

Is food coloring a mixture of substances?

You will need
water
food coloring
a glass jar
filter paper or blotting paper
a pencil
a binder clip
a dropper

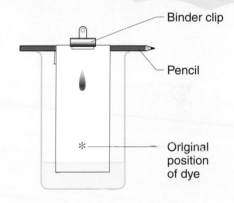

1. Place a little water in the bottom of the jar.
2. Attach a strip of filter paper to the pencil using the clip. The strip should be just long enough to reach the water in the jar when the pencil is laid across the top of the jar.
3. Use the dropper to place a drop of food coloring on the end of the filter paper.
4. Suspend the paper in the jar with the strip just in the water. The pencil will stop it from falling in.

As the water carries the food coloring up through the paper, the different components in it will separate out as they travel through the paper at different rates.

Recycling

An increasing number of the materials we use can now be recycled. Recycling means collecting, processing, and reusing materials instead of throwing them away. Materials that are commonly recycled include paper, glass, and aluminum and steel cans. By recycling we conserve **raw materials** and save the energy that would otherwise be used to make new products. Recycling also saves the space that would have to be used for landfills to deposit all the unwanted garbage. It also helps reduce the pollution that results if garbage is burned.

Recycling materials not only saves raw materials and energy, it also reduces the need for unsightly landfills.

Using recycled materials

Recycling is not a new idea. It has been used in the iron and papermaking industries since the beginning of the twentieth century. Today, recycled materials are used in a variety of products. Aluminum from recycled cans can be used to make new cans and other products for far less energy than would be needed to extract aluminum from **ore.** Recycled paper can be used for insulation as well as in making paper and cardboard. Waste glass can be ground up and melted to make new glass containers. Some plastics can be melted and reshaped to make new plastic products. Motor oil can be recycled and used as industrial fuel oil.

Energy efficiency

Melting down an aluminum can for recycling takes just five percent of the energy that would be needed to make a completely new can.

Sorting out the garbage

Materials for recycling have to be separated out from other waste materials. Glass must first be separated by color before it can be melted and reused. Many local authorities encourage people to do this by providing bottle banks for the disposal of glass, as well as other containers for the collection of paper, cans, clothes, and other recyclable materials. Sorting materials out helps to prevent them from becoming contaminated by unwanted substances and so increases their value.

Separating recyclable materials can also be done at waste-processing plants. Some plants have equipment that separates heavier materials such as metals, glass, and plastic containers from lighter materials that are burned to produce power for the plant. A conveyor belt carries the garbage past **electromagnets** and various mechanical devices that separate out metals and lighter objects from heavier glass. The metals can be sold as scrap or melted down and reused.

Reusing metals

The metals in old cars can be reused. The car body is shredded and large electromagnets are used to remove the iron and steel. Other metals, such as aluminum, copper, zinc, and lead, can be separated by hand. Another method of separation is to use froth flotation. In this process, the metals are placed in a **solution,** the **density** of which can be changed. As the solution is made more or less dense, each metal floats to the surface depending on its own density.

Powerful electromagnets are used in scrap yards to separate out iron and steel.

Glossary

activated carbon charcoal that has been heated to increase its adsorbtive powers

adsorbent solid that holds molecules of a gas or liquid on its surface in a thin film

alloy mixture of two or more metals, or a metal and a non-metal

anode positively charged electrode

atoms tiny particles from which all materials are made; the smallest part of an element that can exist

bonds forces that hold atoms together in molecules

carburetor device in an internal combustion engine for mixing a fine spray of fuel with air

cathode negatively charged electrode

ceramic non-metallic solid that stays hard when heated

chemical reaction reaction that takes place between two or more substances in which energy is given out or taken in and new substances are produced

colloid mixture consisting of tiny particles of one substance evenly scattered through another

compound substance that is made up of atoms of two or more elements

condense to change from a gas into a liquid

conduct to transmit electricity or heat by conduction

density compactness of a substance

dissolve to become incorporated into a liquid and form a solution

distillation way of separating a pure liquid from a mixture

electric current flow of electric charge through a substance

electrodes conductors through which electricity enters or leaves something

electromagnet magnet produced by running an electric current through a coil of wire

electrons negatively-charged particles that are found in all atoms and that are the main carriers of electrical energy

element substance that cannot be broken down into simpler substances by chemical reactions; an element is made up of just one type of atom

evaporation process by which a liquid turns into a vapor without reaching its boiling point

film very thin, flexible sheet of some material; or a thin layer covering a surface

filter device for removing particles from a liquid or gas passed through it

filtrate liquid or gas that has been passed through a filter

fractional distillation method of separating liquids with different boiling points

furnace chamber in which materials can be heated to a very high temperature

graphite form of carbon

insoluble something that cannot be dissolved

ion atom or group of atoms that has an electric charge

kinetic energy energy of movement

leach to remove a soluble substance from a material

mass amount of matter that something contains

microorganisms living things too small to be seen with the naked eye

minerals naturally occurring solid substances; substances obtained by mining

mixture material made of different substances mixed together but not combined chemically

molecule two or more atoms combined together; if the atoms are the same it is an element, if they are different it is a compound

ore rock from which metals can be obtained

particle tiny portion of matter

porous describes a material that has tiny spaces through which liquids or gases can pass

pressure force pushing on a given area

raw materials materials used in the manufacture of something

refine to remove impurities

residue solid material left behind when a liquid or gas is passed through a filter

solute material dissolved in a solvent to form a solution

solution mixture of one substance dissolved in another

solvent part of a solution that dissolves the solute

suspension liquid containing small particles of a solid

vaporize to turn into a vapor

vapor type of gas

volatile liquid that readily turns into a vapor

volume amount of space an object takes up

More Books to Read

Fullick, Ann. *Matter.* Chicago: Heinemann Library, 1999.

Riley, Peter D. *Materials.* Chicago: Heinemann Library, 1998.

Showers, Paul. *Where Does Garbage Go?* New York: HarperCollins Children's Books, 1994.

Index